AF820993

Computational & Experimental Studies on Diaminomaleonitrile Asymmetric Schiff Bases for Fluoride Recognition

Molecular Recognition via Diaminomaleonitrile Asymmetric Schiff Bases

Computational & Experimental Studies on Diaminomaleonitrile Asymmetric Schiff Bases for Fluoride Recognition

Molecular Recognition via Diaminomaleonitrile Asymmetric Schiff Bases

Masood Ayoub Kaloo
Bilal Ahmad Bhat
Mohd Aslam Rather

ELIVA PRESS

ISBN: 978-99949-8-532-6

Cover Design: Eliva Press
Cover Image: Ingimage
Email: info@elivapress.com
Website: www.elivapress.com

This Eliva Press imprint is published by the registered company Eliva Press Global Ltd. part of Eliva Press S.R.L. Publishing Group

The registered company address is: Pope Hennessy Street Level 2, Hennessy Tower Port Louis, Mauritius
Eliva Press S.R.L. Publishing Group address is: Bulevardul Moscova 21, Chisinau, Moldova, Europe

BOOK TITLE:

Computational & Experimental Studies on Diaminomaleonitrile Asymmetric Schiff Bases for Fluoride Recognition

AUTHOR DETAILS:

Dr. Masood Ayoub Kaloo*
Assistant Professor, Department of Chemistry, Govt. Degree College Doda, Jammu & Kashmir, (India)-182202 (Email of Corresponding Author: makandchem@gmail.com)

Mr. Bilal Ahmad Bhat
Research Assistant, Department of Chemistry, Govt. Degree College Doda, Jammu & Kashmir, India-182202

Dr. Mohd Aslam Rather
Assistant Professor, Department of Chemistry, Govt. Degree College Doda, Jammu & Kashmir, India-182202

PREFACE & SUMMARY:

The authors take pleasure in presenting this book ***Computational & Experimental Studies on Diaminomaleonitrile Asymmetric Schiff Bases for Fluoride Recognition***. The entire book contains an updated and latest research conducted in the aspects of the title. The book has been divided into four chapters according to the various types of molecular receptor of interest to be discussed. At the end of each chapter selected and some specific references have also been mentioned therein.

We take this opportunity to express our sincere thanks to all the colleagues and scholars who took pains to send us their valuable suggestions and support for the betterment and improvement of the book.

The authors sincerely hope that the present book will be warmly received by the research scholars and academicians who have keen interest in the field and the growing environmental problems. Authors strongly request the reader's suggestions and valuable inputs for further improvement of this work.

In this book, we have compiled the literature regarding fluoride anion recognition & its mechanism via some novel Diaminomaleonitrile (DAMN) Asymmetric Schiff Bases. Herein we will be restricting ourselves with 2-Amino-((E)-(4-cyanobenzalidine) amino), Naphthyl-DAMN Schiff base and phenyl-DAMN Schiff base derivatives. Such molecular receptor systems possess extensively conjugated framework and belong to a class of organic π-conjugated compounds with electronic donor (D) and acceptor (A) parts connected by single and double bonds. Such D-π-A molecules are push-pull systems and exhibit interesting electronic properties due to intramolecular charge transfer (ICT) processes. These ICT chromophores being highly robust with exceptional light, heat, and weather stability, have been intensively studied as tunable materials for non-linear optics (NLO), optical data storage devices, organic light-emitting diodes (OLED), organic-inorganic hybrid materials, and functional polymers. They are predicted to exhibit novel recognition characteristics for anions like fluoride (F^-) *via* both absorptions as well emission signaling. It has been reported that emission changes are not strong enough to detect F^- through naked-eyes, contrary to the colorimetric output signal, which is prompt and suitable to detect anion even at low concentrations (Micro Molar, μM). The mechanism responsible for naked-eye recognition of F^- was reported to be hydrogen bonding proton transfer from Receptor to F^-. In addition to all this, indirect information about photophysical and photochemical characteristics of the receptor molecule before and after ion recognition was investigated through computational calculations. Experimental observations were correlated theoretically with the help of density functional theory (DFT) and time-dependent density functional theory (TD-DFT). All the theoretical calculations were performed using Gaussian **09** program. The density functional theory and Time-Dependent calculations were run for the excited state and electronic spectra. B3LYP/6-31G+ (*d, p*) and was used to obtain the geometry optimization and IRC parameters of the neutral receptor. The geometry optimization and IRC parameters of the deprotonated form of receptor were also carried out with the same method. To investigate the solvent effect of THF, PCM (polarized continuum model) calculation was performed throughout. In addition, ^{1}H NMR calculations were performed at MPW1PW91/6-311+G (*d, p*) level along with the gauge-independent atomic orbital (GIAO) method. In this work, herein optimized structures and geometry of receptors before and after the anion recognition events, their structural parameters (Bond length, L and Bond angle, Θ), UV–Vis spectra and molecular orbital analysis, Transition state and IRC calculations and conclusions including future aspects are included.

DEDICATED TO PARENTS & FAIMLY

TABLE OF CONTENTS

UNIT 1

Anion Recognition *via* polarized N-H Bonds (Pages: 6-11)

1. Introduction

Molecular recognition events yield information about environmental analytes at molecular level through specific binding of substrate by a molecular receptor. Such binding is regulated by geometrical and electronic complementarily between receptor and analyte. The receptor can be defined as a molecular entity of abiotic origin that interacts with the analyte (anion, cation or a neutral molecule) and offers a unique response (Scheme 1), with a simultaneous signal transduction in the form of photophysical or redox signaling.

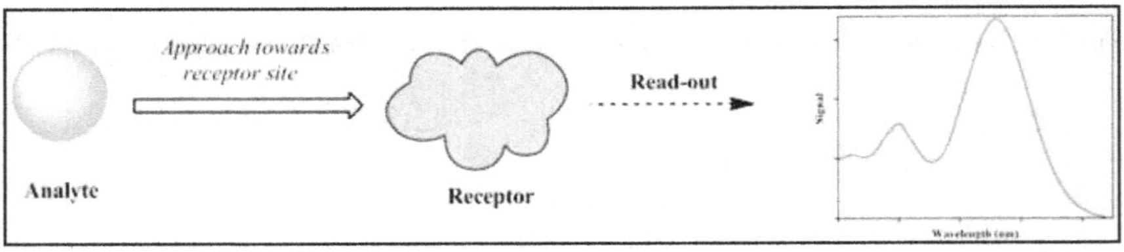

Scheme 1. General schematic representation of the molecular receptor approach.

As far as anion receptors are concerned, the design and synthesis of such receptors capable of binding and sensing anions selectively have received significant attention in recent years because anions play an important role in a wide range of biological, environmental, and chemical processes A. Bianchi *et* all [1]. Anion sensing has been an area of enormous interest to research community owing to the widespread existence of anions in diverse matrices like soil, water, air, food and biosystems.[2] The potential to monitor and chemically characterize such environments is fascinating once we talk of these synthetic tools (*receptors*) achieved through proper combination of organic synthons. The anion and receptor encounter results in the generation of output signals in the form of color, fluorescence, electrochemical, etc which decipher the qualitative and quantitative estimation of the anions in the sample. [3] The use of such low cost, easy to handle, minute, quick time response based molecular platforms have proven to provide the potential alternative towards real-time and convenient monitoring of the diverse environments existing in our day to day arena. Especially, colorimetric and ratiometric chemo sensors Y. Deng *et al* appear to be particularly attractive due to their simplicity, high sensitivity, and high selectivity J.F. Callan *et al* [4-8]. For quantitative analyses, ratiometric chemo sensors have the significant advantage of their dual emission system, which can minimize measurement errors caused by fluctuations of light scattering as well as reagent concentration B. Valeur *et* al [9]. Colorimetric chemo sensors provide an immediate qualitative signal, which allows direct naked-eye detection of anions because of a specific color change of solution upon anion complexation C. Coll *et* al [10-11]. Among the range of biologically important anions, fluoride, the smallest anion, has unique chemical properties. It is of particular interest to recognize and detect it owing to its essential role in a broad range of biological, medical, and chemical processes of osteoporosis, fluorination of drinking water supplies, or even in chemical and nuclear warfare agents K.L. Kirk *et* al [12-15]. However, the chemo sensors for F^- are rather few, and there is a need for good anion sensors with an improved specific response, in particular, with selectivity for F^- in the presence of other anions. Furthermore, it is necessary to understand the unique host-guest interaction of the sensor with F^- and other anions. Different signaling mechanisms have been suggested for F^-, such as photo induced electron transfer (PET) S.K. Kim *et al*, excited state proton transfer (ESPT) X. Peng *et al* and intramolecular charge transfer (ICT) Z. Li *et* al [16-19].

2. Molecules with Polarized N-H fragments

Hydrogen bonding interaction and or proton transfer assets of synthetic molecules in presence of anionic species is pretty fascinating in the field of supramolecular analytical chemistry. Not only amide or urea-based derivatives have appeared in the highlights, rather from last few decades, polarized free amine fragments (-NH_2) have been brought under the study with prompt signaling. Among the various categories of receptors by which molecules sense the anions, utilization of organic moieties with polarized N-H bonds has been is one the most commonly utilized receptor platform in the supramolecular analytical chemistry. In this direction, the arena has got its shift from amide or urea based fragments (Fig. 1a and 1b) to free –NH_2 moiety based molecules possessing polarized N-H bonds.

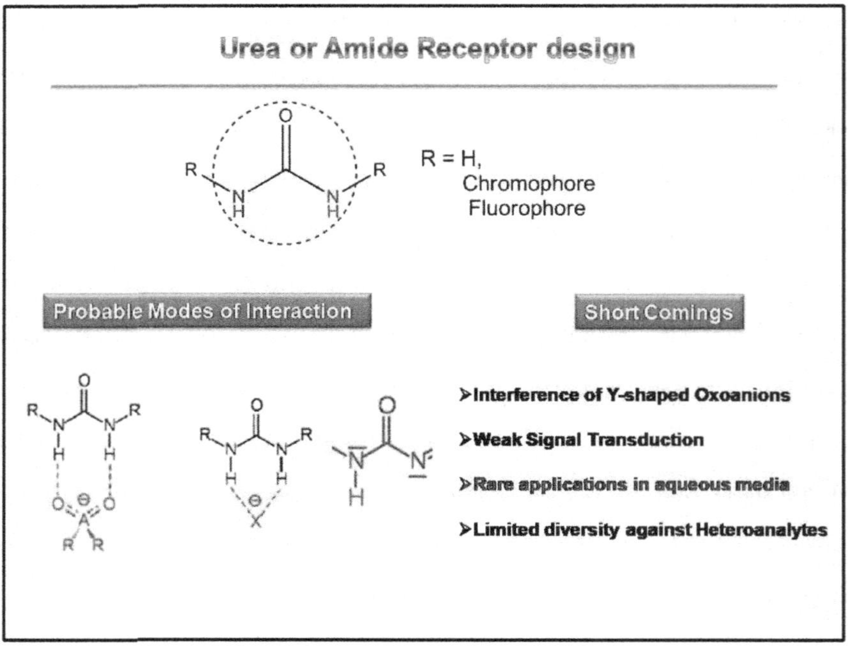

Fig. 1a. Amide or Urea based Molecular Receptor design.

Organic molecules containing N-H bonds wherein hydrogen atom is attached to electron deficient nitrogen center is referred to as polarized N-H bond. Such N-H bonds are behaved as hydrogen atom donors towards anions in polar aprotic solvents (THF, DMSO, CH_3CN, DMF, etc.).[20] Researches with allied fields have utilized various organic moieties which contain such polarized N-H bonds for sensing of anions for diverse applications. This is attributed to the fact such an anion receptor encounter has been most of the times reported with beautiful and astonishing visual display in the form of colour and fluorescence. Not only this, research has extended beyond the existing molecular fragments and scientists started designing new molecules as per the target anion. This has been aided by advanced developments in the synthetic organic chemistry. In this molecular engineering several fields of sciences have clubbed together to attains the specific goals and suffice needs of modern day to day life. Some of important fields include transition metal coordination chemistry, anion coordination chemistry, supramolecular chemistry, analytical and bio analytical sciences as well as materials sciences.[21]

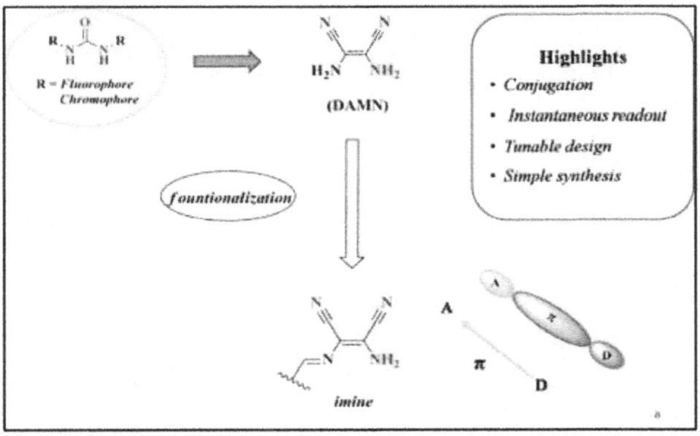

Fig. 1b. The scheme showing shift of molecular receptor design from Urea based receptor design towards free $-NH_2$ based Receptor design.

For the generation of molecules with polarized N-H bonds, a range of molecular fragments have been reported for studying anion recognition chemistry. In this regard, molecules with functionalities like, ureas, amides, thioureas, ammonium, imidazole and imidazolium systems have been reported.[22] The interactions of anions under the aprotic solvent conditions have resulted in the modulation of electronic properties in the systems which are accompanied by output signals (absorption, emission, electrochemical, etc.). However such systems interact with anions via hydrogen bond donation or in extreme cases accompanied by proton transfer from molecule to anion. The events result in the enhancement of electron density on the nitrogen atoms. The concentration of negative charge density on the nitrogen modulated various processes like photo induced electron transfer (PET), Charge transfer (CT), conjugation, etc (Fig. 2).[23]

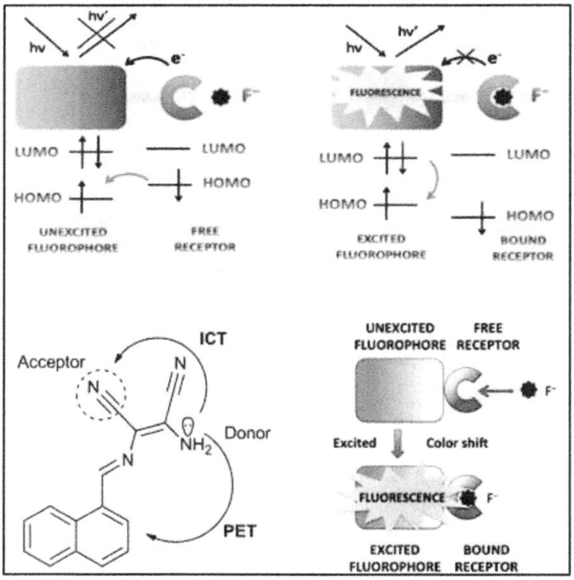

Fig. 2. PET and CT Processes in molecular Receptors.

The modulation of such phenomenon produces the basis for locating the anion in any matrix. The changes could be qualitative as well as quantitative. Not only this, the anion signaling events are very fast (within few seconds) and most of the times reversible. Thus signifying the use of these polarized N-H bond based molecules for monitoring of such anionic analytes with an ease. Despite all this, certain short comings have been noticed in such class of molecules with N-H fragments. Molecular scaffolds suffer from multi-anion interference due to Y-shaped oxoanions through hydrogen bonds. The molecular structure requires tagging of markers due to which cost of the material increases. Besides, some of the events ended up with delayed response.[24]

3. DAMN-Schiff base Molecular Receptors

In order to overcome such shortcomings, recently a new strategy wherein polarized N-H bonds have been developed from free $-NH_2$ fragments and suitable for anion recognition.[25] The novel strategy has been usually achieved by an easy to obtain asymmetric Schiff base condensation reaction between aldehyde with the diaminomelonitrile (DAMN). The product obtained produced molecules with free $-NH_2$ fragments responsible for anion sensing via hydrogen bonding or proton transfer mechanism (Fig 3). The details of anion recognition with discussed in coming units of this book in detail. The realm of such anion polarization was found not only suitable to detect anion in organic solvents but suitable enough to sense anions under aqueous environment. The mechanism of action is attributed to the fact that the free $-NH_2$ group has been set as electron donor (D) at one end of the molecule with respect to some electron acceptor (A) at other end of the molecule. The D and A fragments were put in strong electronic communication via *pi* (π) bonds. This pull-push type of molecular architecture results in the flow of electronic charges away from nitrogen atom of free $-NH_2$, hence N-H bond becomes polarized. During the encounter between the anion and the D moiety of molecule which act as receptor site as well, hydrogen bond formation or proton transfer from $-NH_2$ to anion results the enhancement of electron density on the **D** site. This promotes a strong intramolecular charge transfer (ICT) across the system hence a prompt and strong readout signaling comes out. Such characteristic features of the newly designed polarized receptor design have overcome most of the shortcomings associated with the earlier receptor based upon amide or urea based fragments.

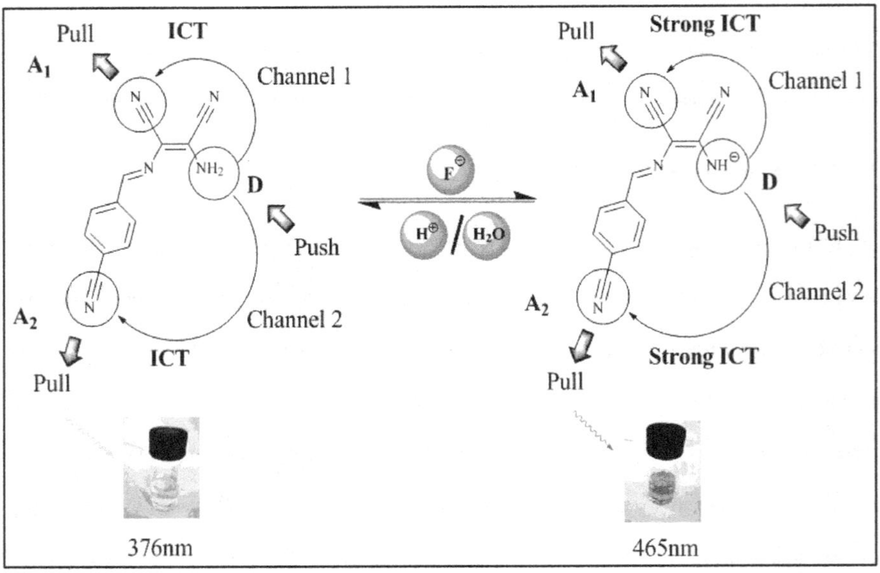

Fig. 3. DAMN Molecular Receptor Design.

4. Conclusions

In conclusions, I should say that the birth of anion recognition via polarized N-H fragments got from natural biological systems like amide bonds in proteins which were embedded by nature and involved in hydrogen bonding interaction. From last five decades researchers have utilized this fact and designed the mimics of this biological architecture in the labs and later utilized for the anon sensing applications in vitro and vivo. Starting from urea or amide based polarized N-H bonds; the field has now got extended to free $-NH_2$ fragments. In general, it has been established that the molecular system recognizes F^- selectively in the presence of different interfering species and displays pronounced colorimetric changes and fluorescence 'turn-on' response as a function of concentration. The recognition signal occurs as a result of ICT enhancement due to intermolecular proton transfer from NH_2 to F^-. Receptors exhibits excellent selectivity and detect a fluoride ion at micro and nanomolar concentrations. The core unsubstituted receptor design is an effective approach to develop new colorimetric and fluorescent anion recognition systems for various charged species. We believe this label-free receptor design based on DMN Schiff bases has a potential future to replace the old, sophisticated and tedious ones from a well known and diverse class of amide (–CONH–) functions. Future looks bright for chemistry and related scientists to sink into this area of this research and expose the futuristic applications.

5. References

1. Bianchi A, Bowman JK, Garcia EE Supramolecular Chemistry of Anions. Wiley-VCH: New York, NY, USA, 1997.
2. P.A. Gale, Chem. Soc. Rev. 2010, 39, 3746-3771
3. D. Esteban-Gomez, L. Fabbrizzi and M. Licchilli, J. Org. Chem. 2005, 70, 5717-5720
4. Deng Y, Chen Y, Cao D, et al. A cationic triarylborane as water-tolerant Fluorescent chemosensor for fluoride anions. Sensors Actuator B-Chem. 2010; 149:165-169.
5. Li JQ, Li XY. Multichannel photo induced intramolecular electron-transfer excitations in a bis-naphthalimide spermine conjugate by time-dependent density functional theory. Phys Chem A 2007; 111:1306-13068.
6. Hagihara S, Tanaka H, Matile S. Boronic acid converters for reactive hydrazide amplifiers: Polyphenol sensing in green tea with synthetic pores. J Am Chem Soc 2008; 130:5656-5657.
7. Yoon J, Kim SK, Singh, et al. Imidazolium receptors for the recognition of anions. Chem Soc Rev 2006; 35: 355-360.
8. Gale PA. Structural and molecular recognition studies with acyclic anion receptors. Acc Chem Res 2006; 39:465-475.
9. Callan JF, de Silva AP, Magri DC. Luminescent sensors and switches in the early 21st century. Tetrahedron 2005; 61: 8551-8588.
10. Martinez M R, Sancanon F. Fluorogenic and chromogenic chemosensors and reagents for anions. Chem Rev 2003; 103:4419-4476.
11. Valeur B Molecular Fluorescence Wiley-VCH: Weinhein, Germany, 2002.

12. Coll C, Marínez MR, Dolores MM, et al. A simple approach for the selective and sensitive colorimetric detection of anionic surfactants in water. Angew Chem Int Ed 2007; 46; 1675-1678.

13. Evans LS, Gale PA, Light ME, et al. Anion binding vs deprotonation in colorimetric pyrrolylamido thiourea based anion sensors. Chem Commun 2006; 55:965-967.

14. Kirk KL. Biochemistry of the Halogens and Inorganic Halides; Plenum Press: New York, NY, USA, 1991.

15. Kleerekoper M. The role of calcitonin in the prevention of osteoporosis. Endocrin Metab Clin North Am. 1998; 27:414-418.

16. Zhang SW, Swager TM. Fluorescent detection of chemical warfare agents: Functional group specific ratiometric chemosensors. J Am Chem Soc 2003; 125:3420-3421.

17. Kim SK, Yoon J. A new fluorescent PET chemosensor for fluoride ions. Chem Commun 2002; 33:770-771.

18. Peng X, Wu Y, Fan J, et al. Colorimetric and ratiometric fluorescence sensing of fluoride: Tuning selectivity in proton transfer. J Org Chem 2005; 70:10524-10531.

19. Li Z and Zhang J An efficient theoretical study on host-guest interactions of a fluoride chemosensor with F^-, Cl^-, and Br^-. Chem Phys 2006; 331:159-163.

20. Thiagarajan V and Ramamurthy P. Specific optical signaling of anions via intramolecular charge transfer pathway based on acridinedione fluorophore. J Lumin 2007; 126:886-892.

21. P. A. Gale, E.N.W. Howe and X. Wu, Chem. 2016, 1, 352-422.

22. V. Amendola, D. Esteban-Gomez and L. Fabbrizzi and M. Licchilli, Acc. Chem. Res. 2006, 39, 343-353.

23. M. A. Kaloo and J. Sankar, New. J. Chem. 2014, 48, 923-926. V. Amendola, L. Fabbrizzi and L. Mosca, Chem. Soc. Rev. 2010, 39, 3889.

24. B. Kumar, M. A. Kaloo, A. R. Sekhar and J. Sankar, Dalton. Trans. 2014, 43, 16164-16168.

25. M. A. Kaloo and J. Sankar, Analyst. 2013, 238, 4760-4763.

Unit-2

2-amino-((E)-(4-cyanobenzalidine) amino) maleonitrile (Pages: 12-26)

1. Molecular Receptor of Interest & Synthesis

For the recognition of fluoride anion, an unprecedented molecular anion receptor with polarized N-H bond in the free NH_2 functionality, 2-Amino-((E)-(4-cyanobenzalidine) amino) maleonitrile (named **1** in Fig. 1) was reported by Sankar *et al.*[1]. The molecular system with a highly extensively conjugated framework belongs to a class of organic π- conjugated compounds with electronic donor (D) and acceptor (A) parts connected by single and double bonds[2-11].

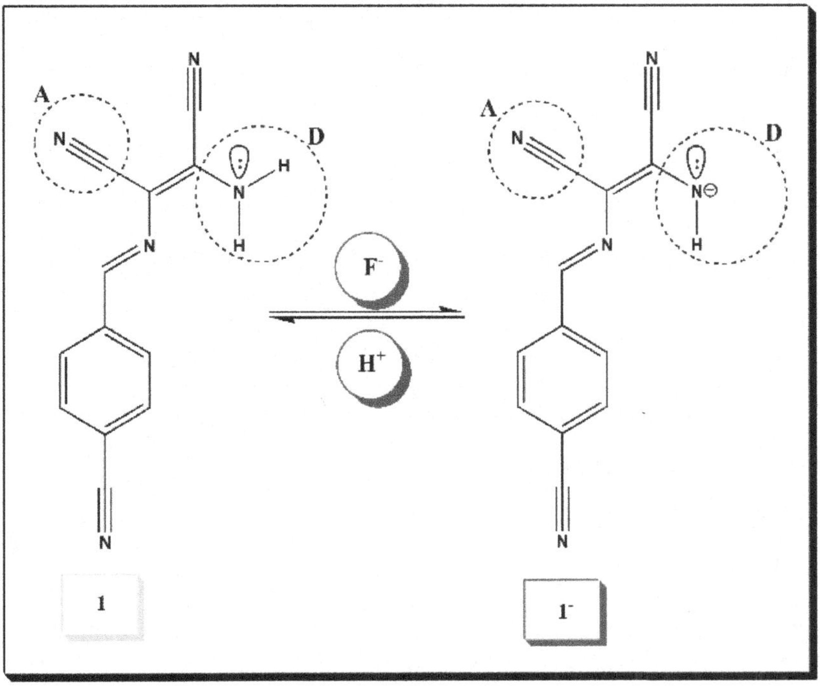

Fig. 1. Proposed mechanism of F^- interaction with the molecular receptor (1) leading to the formation of deprotonated receptor 1^-, and its reversal in the presence of acid (H^+).

Such D-π-A molecules are push-pull systems and exhibit interesting electronic properties due to intramolecular charge transfer (ICT) processes.[12-19] These ICT chromophores being highly robust with exceptional light, heat, and weather stability, have been intensively studied as tunable materials for non-linear optics (NLO), optical data storage devices, organic light-emitting diodes (OLED), organic-inorganic hybrid materials, and functional polymers[20-24]. The novel receptor (Fig. 1) could recognize F^- *via* both absorptions as well emission readout. However, emission changes were not strong enough to detect F^- through naked-eyes, contrary to the colorimetric output signal, which was prompt and suitable to detect anion even at low concentrations (20-50 μM). The mechanism responsible for naked-eye recognition of F^- was reported to be proton transfer from **1** to F^-.

To the stirring solution of 2-equivalents of aldehyde (para substituted benzaldehydes) in distilled water (**Scheme 1**). Add 2-3 drops of conc. HCl, followed by drop wise addition of methanolic solution of 1 equivalents of Diaminomalenonitrile (DMN). The resulted imine solidifies within a minute, was filtered, washed with water and recrystalized with ethanol.

Characterization of the achieved compound was done with the help of standard spectroscopic techniques like NMR and HR MS.

1. Colour: Yellow colour,
2. Yield: 95%
3. ^1H NMR- (500MHz; d6-DMSO):

δ 8.38 (s, 1H), 8.36 (s, 2H), 8.31 (d, *J* = 5Hz, 2H), 8.33 (d, *J* = 5Hz, 2H). 13C NMR-(500MHz; d6-DMSO): δ 152.68, 149.00, 141.74, 130.22, 129.07, 124.20, 114.58, 113.97, 102.36.

4. MS (HRMS, negative mode) found 241.061. C11H7N5O2 requires 241.060.

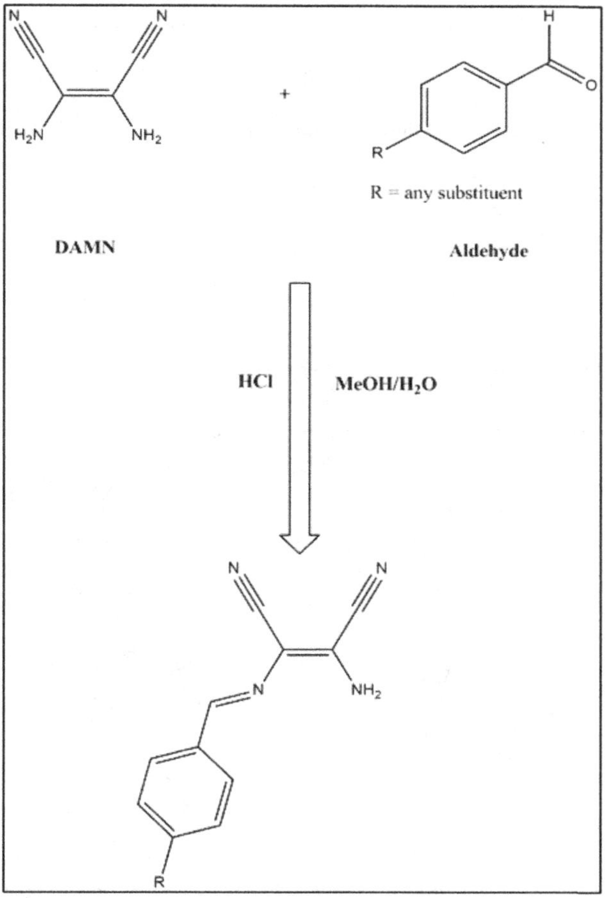

Scheme 1. Scheme for synthesis of Receptor **1**.

2. Computational Methods

All calculations were performed using Gaussian 09 program[25]. The hybrid density functional theory and Time-Dependent DFT calculations were run for the excited state and electronic spectra. B3LYP/6-31G+ (d, p) was used to obtain the geometry optimization and IRC parameters of the **1** as it is moderate and suitable for such organic molecular structures [26]. The geometry optimization and IRC parameters of the deprotonated form of receptor $\mathbf{1}^-$ were also carried out with the same method. To investigate the solvent effect of THF, PCM (polarized continuum model) calculation was performed throughout the steps[27-28]. In addition, ^1H NMR calculations were performed at MPW1PW91/6-311+G (*d,p*) level along with the gauge-independent atomic orbital (GIAO) method.

3. Optimized Structures

The optimized structures of 1 and deprotonated form of the receptor, 1- are shown in Fig. 2.

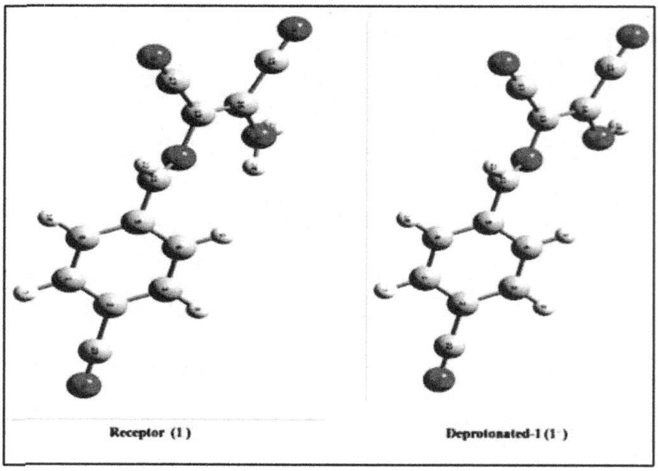

Fig 2. Optimized structures of receptors **1** and $\mathbf{1}^-$.

The ground state (S0) ionization energies (I. Es) of 1 and 1^- are found to be -735.17 and -830.26 A.U., respectively. The bond distance between N19–H20 and N19-H21 in 1 ground state optimized structure and exciting structure seem to be equidistant, i.e., 0.99 Å. In the ground state of 1-, an increase in the bond distance between hydrogen and nitrogen was found, i.e., 1.033 Å, which attributes to the increase in the electronic density of nitrogen. The bond C-N (NH2) bond distance is equal to (1.47 Å) in receptor 1. However, in the case of 1^-, it decreases to 1.36 Å, which attributes to higher bond order between C-N. The important structural parameters (bond length and bond angles) of **1** and **1**$^-$ are further summarized in Table 1.

Table 1. The structural parameters (bond length, L and bond angle, Θ) in the optimized geometry of **1** and **1**$^-$.

C-N (L)	**C=N (L)**	**C=C (L)**	**Ph-C≡N (L)**	**N-H (L)**	**C-N-C (Θ)**	**N-C-H (Θ)**
11.47	1.29	1.316	1.16	0.99	125.7	120.9
11.36	1.29	1.416	1.16	1.03	120.1	120.40

4. UV-Vis Spectra & Molecular Orbital Analysis

Experimental studies with UV-Vis spectra of 1 and 1^- revealed a red shift of its characteristic absorption band (λ_{max}) from 376 to 465 nm[29]. Such a spectral shift ($\Delta\lambda$) of 89 nm was proposed to occur as a result of proton transfer signaling (PTS) from amine hydrogen (NH–H) of 1 to the fluoride. Due to the strong basic nature of F^- in polar aprotic solvents, the interaction of F^- results in the formation of negatively charged molecular species (named herein 1^- in Fig. 1). The concentration of negative charge density on the amine nitrogen results in the enhancement of the electronic push-pull effect between the donor (D) amine moiety and acceptor (A) nitrile (CN) functionality. The facilitation of such D-π-A interaction is responsible for visual and prompt colorimetric sensing of F^-. Similar absorption patterns were observed through the theoretical investigations (Figure 3a and 3b).

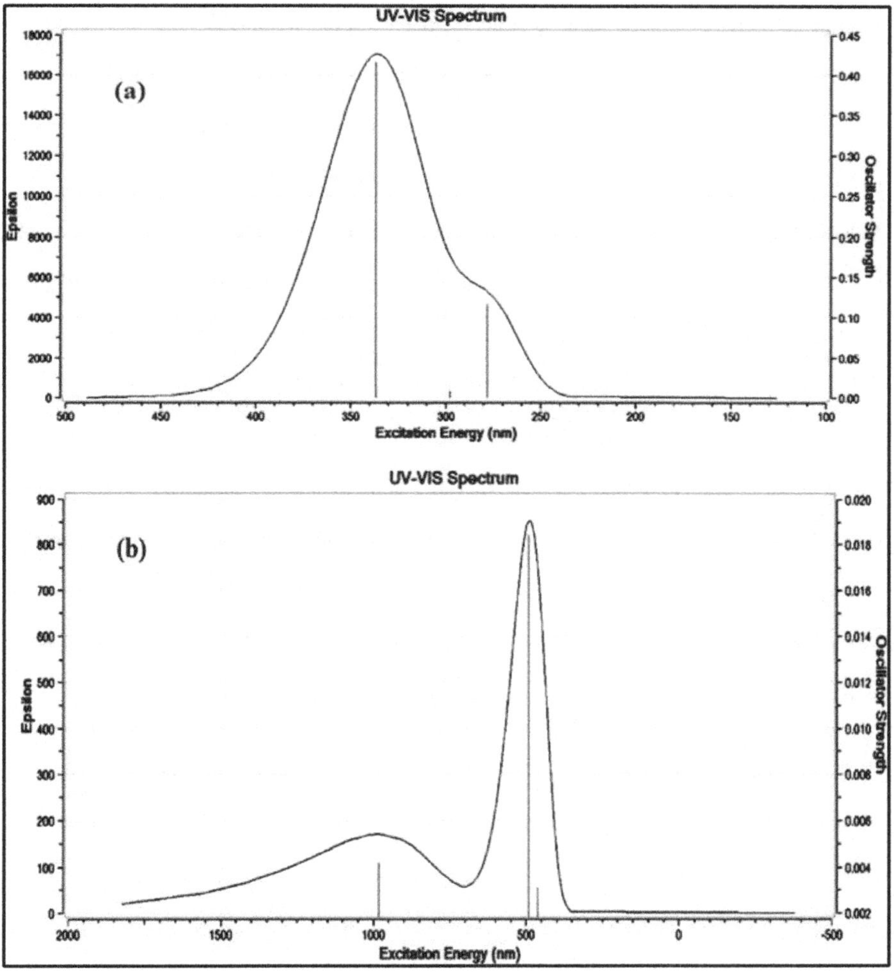

Fig. 3. (a) and (b) Simulated absorption spectra (UV-Vis) of receptor **1** in THF and (b) simulated absorptionspectra (UV-Vis) of the deprotonated form of the receptor, **1**$^-$ in THF.

In order to investigate these photophysical characteristics of receptors in the presence and absence of fluoride anion (1 to 1^-), a molecular excitation study of both these molecules was carried out (Figs. 4a and 4b). Simulated spectra of receptor 1 showed an intense absorption peak at 336 nm as a dominant π-π* type transition with the oscillator strength (*f*) of 0.4162, which is from the highest occupied molecular orbital (HOMO) to the lowest unoccupied molecular orbital (LUMO). The calculated wavelength of absorption maximum corresponds to the energy gap (ΔE) of 4.014 eV. The *pi*-electron density in the HOMO delocalized both through phenyl moiety and an amine group. In the local transition, these delocalized electrons (HOMO) were transited only to phenyl moiety (LUMO). However, in $\mathbf{1}^-$, an intense absorption peak at 493 nm was obtained, resulting from dominant π-π* type transition from the highest occupied molecular orbital (HOMO) to the lowest unoccupied molecular orbital (LUMO) with 2.342 eV energy gap. The corresponding oscillator strength was calculated to be $f = 0.0184$. Thus, the decrease in the energy gap between HOMO and LUMO of receptor in the presence of F^- realizes a strong spectral shift (red shift), and hence offering naked-eye optical detection of anions.

In addition to the HOMO to LUMO transition in both the neutral and deprotonated receptor molecule, other transitions with minor contributions towards photophysical behavior were also calculated (Fig. 3a). The second dominant transition in **1** occurs from HOMO-1 to LUMO with ΔE = -3.837 eV. Other fewer energy transitions include HOMO-2, LUMO, and HOMO to LUMO +1 with ΔE of 3.865 eV and 3.216 eV, respectively. Similarly, HOMO-1 to LUMO, HOMO-2 to LUMO, HOMO to LUMO +1 transition in the deprotonated form of the **1** have been shown in Fig. 3b.

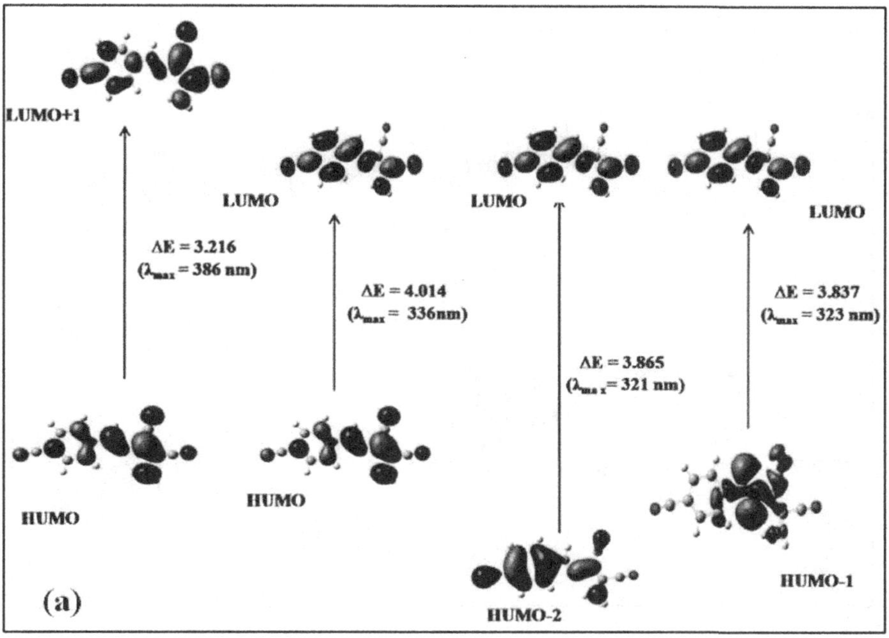

Fig. 4(a) Frontier molecular orbitals of the receptor, **1** before in the absence of F^-. The energy gap (ΔE) between corresponding transitions and their maximum absorption wavelengths are mentioned herein.

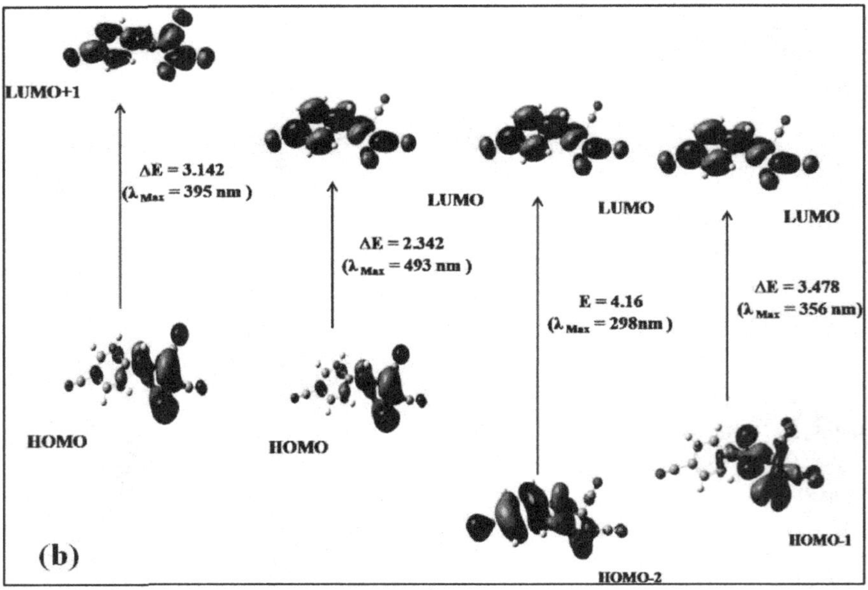

Fig. 4. (b) Frontier molecular orbital's of receptor in the presence of F^- ($\mathbf{1}^-$). The energy gap (ΔE) between corresponding transitions and their maximum absorption wavelengths are mentioned herein.

5. ^{1}H NMR Spectral Studies

To reveal the deprotonation mechanism in the fluoride ion sensing, ^{1}H NMR calculations were performed at the MPW1PW91/6-311+G (*d,p*) level and with the gauge-independent atomic orbital (GIAO) method for **1** and **1**$^-$. The calculated results were further compared with experimental data. For **1**, the 1H NMR signals at 8.27 ppm are assigned to the equivalent protons of the NH2. Whereas in the case of 1^-, the N-H signal shifted to 22.23 ppm, whereas the phenyl-C-H protons also shifted to up-field. The aromatic proton signals also shifted to up field, implicating

PT and induced charge transfer (ICT). The calculated NMR spectra (Fig. 5) of 1 and 1^- agree quite well with experimental observations[30]. Thus, it is evidenced that in the presence of fluoride ion, the hydrogen atom is transferred from **1** to the fluoride anion to form a deprotonated structure, 1^-. This transfer of proton results, in turn, results in the enhancement of ICT across the receptor. The important data is tabulated below in Table 2.

Table 2. ^{1}H NMR data of **1** and **1**$^-$.

Proton	**Receptor (1) (Shift)**	**1- (Shift)**
H_7	8.06	8.02
H_8	8.01	8.00
H_9	8.00	8.23
H10	7.89	7.88
H20	4.45 (NH_2)	-
H21	4.45 (NH_2)	-
H_{22}	8.67 (C-H)	8.75

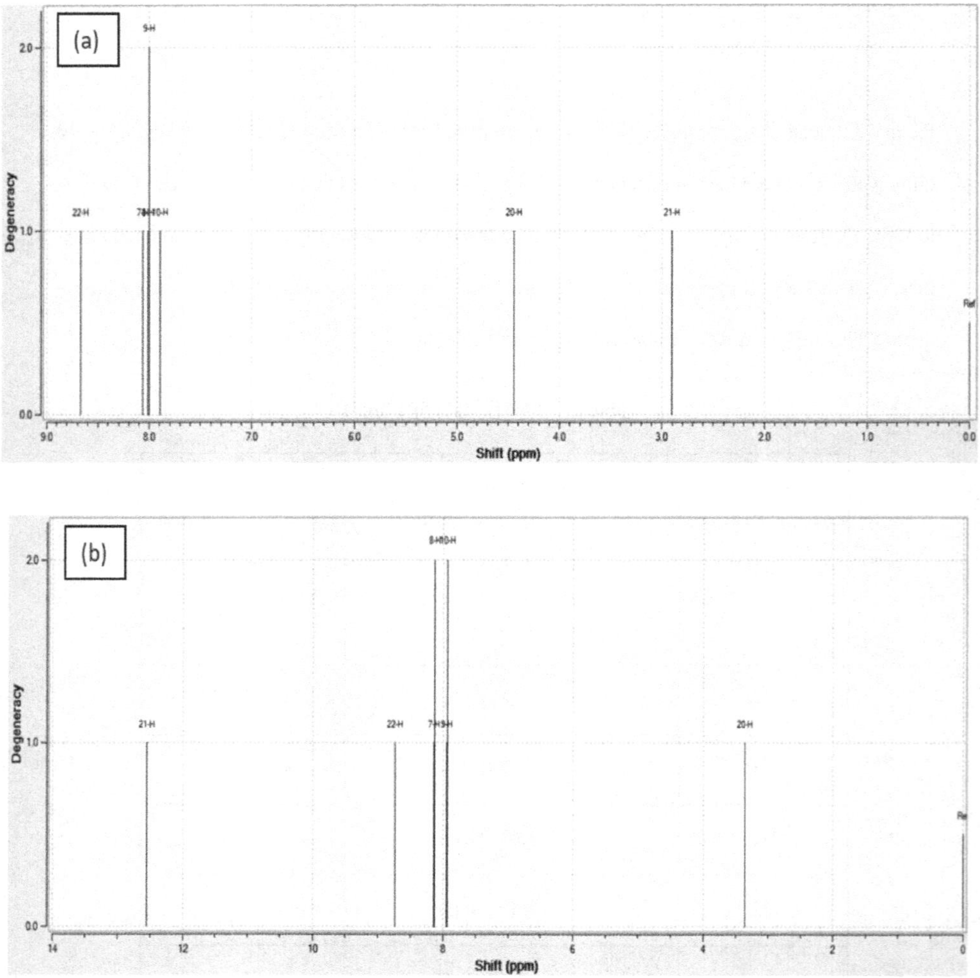

Figure. 5. (a) Theoretically calculated ^{1}H NMR of Receptor **1** and (b) deprotonated form of the receptor, **1**$^-$.

6. Transition State & IRC Calculations

In order to capture the dynamic feature of the proton transfer reaction, Gibb's free energy in the solution was calculated. The ground state Gibb's free energy of the **1** and **1**$^-$ were found to be -1.171 kcal/mol and -1.323 kcal/mol, respectively. Hence after the interaction between the receptor **1** and F^-, the total change in Gibbs free energy is-0.152 kcal/mol. The negative value indicates the feasibility of the sensing process in thermodynamic aspects. The bond angle and bond length of various bonds of **1**$^-$ resonate from 118.3-120 Å and 1.395-1.415 Å. As shown in Table 1, C=C bond length increases and C-N bond length decreases, respectively. This may be due to the single bond and double bond character of these bonds in the receptor in the presence of fluoride anion. While the bond angle decreases between C13, N12, H21, and similarly between N12, C11, H22 atoms. All these changes may be attributed to the enhancement of conjugation across the donor- acceptor unit. The IRC parameters, i.e., total energy, bond energy, and bond angle for **1**$^-$ are shown below in Fig. 6.

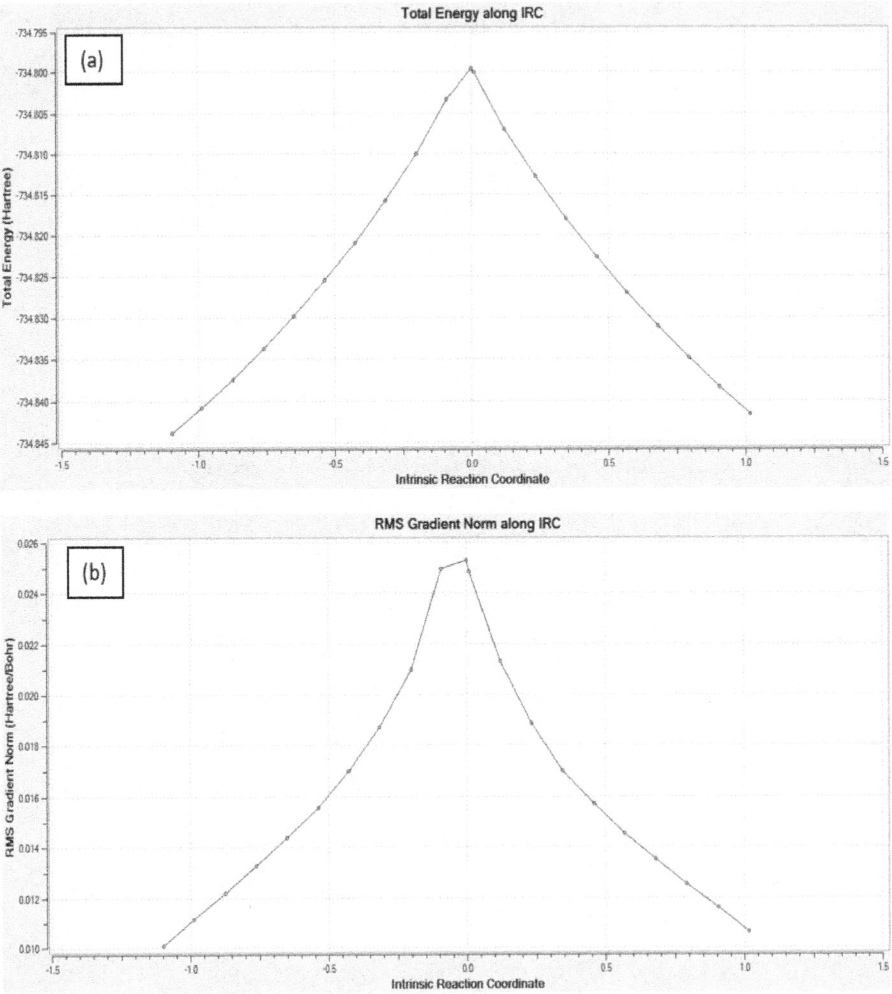

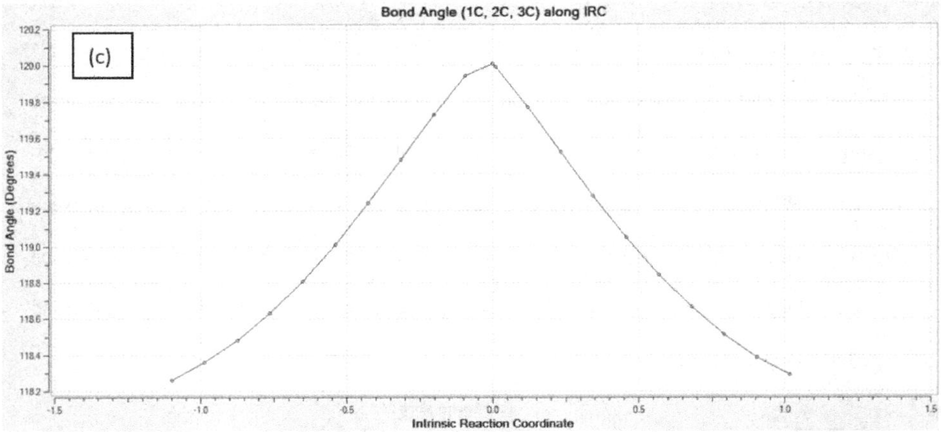

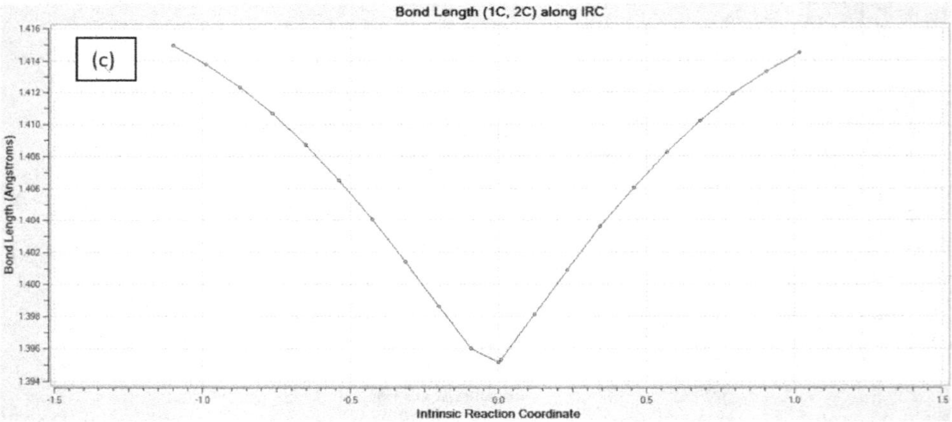

Fig. 6. (a) Theoretically calculated total energy, (b) Gibbs energy, (c) bond angle, and (d) bond length of the deprotonated for the receptor $\mathbf{1}^{-}$.

7. Conclusions

In conclusion, we have investigated the sensing mechanism of an anion receptor 1 with the help of quantum computational (DFT/TD-DFT) methods and mainly discussed the 1H NMR and UV–vis spectra. A receptor molecule possessing D-π-A backbone was established to perform prompt colorimetric sensing of F−through PTS signaling mechanism. The sensing of fluoride anion is attributed to their higher basicity or/and hydrogen bonding tendency under polar aprotic environment (THF). The FMOs analysis indicates intermolecular charge transfer (ICT) enhancement through deprotonation. Fluoride-induced deprotonation decreases the HOMO-LUMO energy gap, which correlates to the bathochromic shift in UV-Vis spectra. The PTS mechanism was further probed through calculated 1H NMR. The experimentally calculated results were in good agreement with the theoretical observations

8. References

1. M. A. Kaloo and J. Sankar, Analyst 138, 3 (2013). https://doi.org/10.1039/C3AN00942D.
2. M. Fokkens, T. Schrader, and F. G. Klarner, J. Am. Chem. Soc. **127**, 41 (2005). https://doi.org/10.1021/ja052806a
3. J. W. Jones and H. W. Gibson, J. Am. Chem. Soc. **125**, 12 (2003). https://doi.org/10.1021/ja0689334
4. J. W. Jones, L. N. Zakharov, A. L. Rheingold, and H. W. Gibson, J. Am. Chem. Soc. **124**, 10 (2002). https://dx.doi.org/10.1021/ar500046r
5. P. A. Gale, Chem. Soc. Rev. **39**, 3 (2010). https://doi.org/10.1039/c001871f
6. J. Liu, Y. Liu, Q. Liu, C. Li, L. Sun, and F. Li, J. Am. Chem. Soc. **133**, 12 (2011). https://doi.org/10.1021/ja109659k
7. T. T. Christison and J. S. Rohrer, J. Chromatogr. A, **1155**, 1 (2007). https://doi.org/10.1016/j.chroma.2007.02.083
8. P. D. Beer and P. A. Gale, Angew. Chem. Int. Ed. **40** (2001). https://doi.org/10.1002/1521-3773.
9. C. Perez-Casas and A. K. Yatsimirsky, J. Org. Chem.**73**, 2 (2008). https://doi.org/10.1007/s10847-010-9813-5
10. F. Y. Wu, Z. C. Wen, N. Zhou, Y. F. Zhao, and Y. B. Jiang, Org. Lett. **4**, 5 (2002). https://doi.org/10.1039/B926320A
11. J. M. Linares, D. Powell, and J. K. Bowman, Coord. Chem. Rev. **240**, 13 (2003). https://doi.org/10.5155/eurjchem.2.3.410-415.189

12. J. M. Kang, H. S. Kim, and D. O. Jang, Tetrahedron Lett. 46, 7 (2005). https://doi.org/10.1002/cjoc.200790115

13. K. Chellappan, N. J. Singh, I. C. Hwang, J. W. Lee, and K. S. Kim, Angew Chem. Int. Ed. 44,

14. 10 (2005). https://doi.org/10.1063/1.2147283

15. N. Ahmed, V. Suresh, B. Shirinfar, I. Geronimo, A. Bist, I. C. Hwang, and K. S. Kim, Org. Biomol. Chem. 10, 3 (2012). https://doi.org/10.1007/s12274-020-2718-8

16. V. Amendola, L. Fabbrizzi, and L. Mosca, Chem. Soc. Rev. 39, 9 (2010). https://doi.org/10.1039/B822552B

17. J. R. Farley, J. E. Wergedal, and D. J. Baylink, Sci. 222, 3 (1983). https://doi.org/10.1007/BF02555092

18. J. M. You, H. Jeong, H. Seo, and S. Jeon, Sens. Actuators B Chem. 146, 160 (2010).

19. A. Aggeli, M. Bell, L. M. Carrick, C. W. G Fishwick, R. Harding, P. J. Mawer, S. E. Radford, A. E. Strong, and N. Boden, J. Am. Chem. Soc. 125, 3 (2003). https://doi.org/10.1021/ja021047i

20. I. H. A. Badr and M. E. Meyerhoff, J. Am. Chem. Soc. 127, 15 (2005). https://doi.org/10.1021/ja0500153

21. M. Melaimi and F. O. P. Gabbai, J. Am. Chem. Soc. 127, 27 (2005). https://doi.org/10.1021/ja053058s

22. D. Jacquemin and C. Adamo, Computational Molecular Electronic Spectroscopy with TD-DFT (Springer, Berlin, 2015). https://doi.org/10.1007/128_2015_638

23. A. D. Laurent, C. Adamo, and D. Jacquemin, Phys. Chem. Chem. Phys. 16, 3 (2014). https://doi.org/10.1039/C3CP55336A

24. J. Hossen and T. K. Pal, J. Sci. Res. 13, 2 (2021). https://doi.org/10.3329/jsr.v13i2.51731

24. G. Gayatri, J. Sci. Res. 13, 1 (2021). https://dx.doi.org/10.3329/jsr.v13il.48371

25. B. Moorell, A. Charaf-Eddin, A. Planchat, C. Adamo, J. Autschbach, and D. Jacquemin, J. Chem. Theory Comput. 8, 7 (2014). https://doi.org/10.1021/ct300326f

26. D.Yang, P. Li, R. Zheng, Y. Wang, and J. Lv, Theor. Chem. Acc. 135 (2016). https://doi.org/10.1007/s00214-015-1802-8

27. M. J. Frisch, et al., Gaussian 09. Revision A 02 Gaussian Inc. Wallingford CT (2009).

28. I. S. K. Kerkines and I. D. Petsalakis, J. Phys. Chem. A 115, 6 (2011). https://doi.org/10.1021/jp1088433

29. M. Takeuchi, T. Shoya, and T. M. Swager, Angew Chem. Int. Ed. 40, 3372 (2001). https://doi.org/10.1002/1521-3773

30. Y. Qu , J. Hua, and H. Tian, Org. Lett. 12, 4 (2010). https://doi.org/10.1021/ol9029066

Unit-3

2-amino-3-((E)-(napthalen-1-ylmethylene) amino) maleonitrile (Pages: 27-34)

1. Molecular Receptor of Interest & Synthesis

Molecular Receptor, 2-amino-3-((E)-(napthalen-1-ylmethylene) amino) maleonitrile (R) was proposed by Masood *et al.*[1] with the endeavor to develop a simple molecular arithmetic machine sufficing above criteria, They looked forward to such π-conjugated organic chromophores with extended charge transfer (ICT) and electron transfer (PET) properties.[1-11] The later has been intensively studied for non-linear optics (NLO), optical data storage devices and organic light emitting diodes (OLED). These systems were reported to usually exhibit very high sensitivity to the external perturbations in the form of anions and cations.[12-15] With these considerations in mind, 2-amino-3-((E)-(napthalen-1- ylmethylene) amino) maleonitrile (**R**) was brought under study (Fig. 1a). It was obtained by a new and simplest ever synthetic procedure. It involves equimolar addition of methanolic solution of Diaminomalenonitrile (DMN) to a stirring solution of 1-napthaldehyde in water, containing a catalytic amount of concentrated HCl. A yellow colored solid got precipitated immediately. Synthetic procedure involves drop wise addition of DMN (50 mg, 0.46 mmol; methanolic solution) to same equivalent of 1-napthaldehyde (71mg, 0.46mmol) in distilled water containing 1-2 drops of conc. HCl (**Scheme 1**). An immediately solidified imine (R), was filtered, recrystalization with ethanol/water mixture (v/v 50 %).

The compound obtained was finally characterized by NMR (^{1}H NMR, ^{13}C NMR) and HR-MS. R (Yellow colored, 95 % yield);

1H NMR- (500MHz; d6-DMSO): δ 8.97 (s, 1H), 8.01 (s, 2H), 8.76 (d, J = 10Hz, 1H), 8.42 (d, J = 10 Hz, 2H), 8.10 (d, J = 5 Hz, 1H), 8.04 (d, J = 5 Hz, 2H), 7.70 (t, J = 15Hz, 1H),7.65 (t, J = 10

Hz, 1H), 7.62 (t, J = 10Hz, 1H), 13C NMR- (500 MHz; d6-DMSO): δ 153.95, 133.87, 132.66, 131.20, 130.60, 129.70, 129.33, 128.39, 127.34, 126.85, 125.99, 123.79, 11.96, 114.31, 104.04.

HR-MS (negative mode) found 248.088 for C15H10N4 and Calcd. 248.106

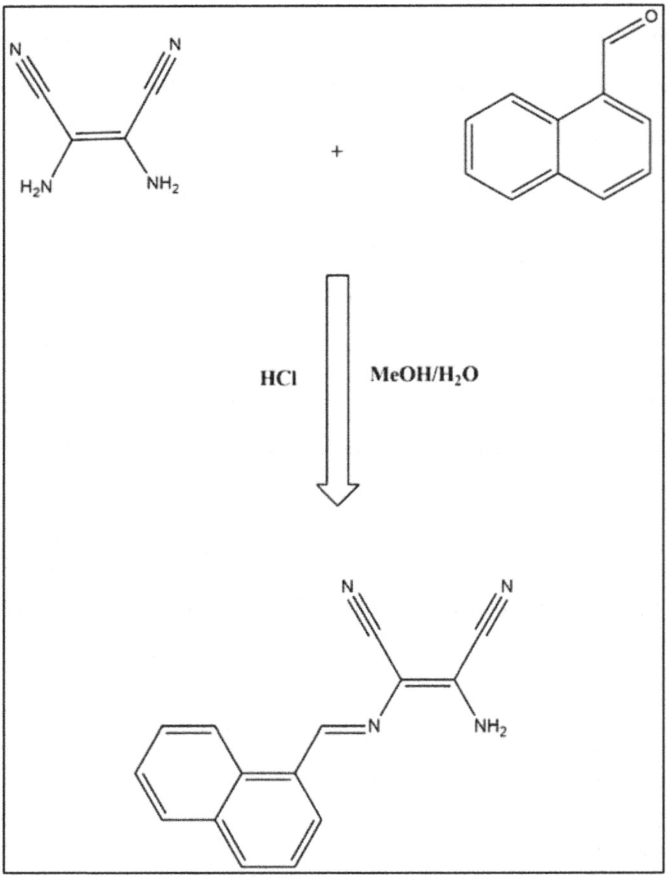

Scheme 1. Scheme for synthesis of Receptor **R**.

2. Computational Method

In the present work, one of the hybrid methods; Lee-Yang-Parr correlation function (B3LYP) was carried out by basis sets 6-311++G (*d,p*) using GAUSSIAN 09W program. In DFT methods; Becke's hybrid function combined with the B3LYP function predict the best results of Mullikan charge density on N and H of amine (NH_2) group, and absorption spectra of receptor (R) and its deprotonated form (R^-). The theoretical result of electronic absorption transitions was compared with the experimental.

3. Mullikan Charge distribution analysis

The Mullikan charge on the atoms is used to understand the charge distribution on the chemical bonding. The analysis facilitates positive and negative regions in the molecular space, at which the protons and electrons concentrate. Thus chemically significant regions and bonds can be identified. This also gives the description of the possible mechanism in the form of electrophilic, nucleophilic substitutions, hydrogen bonding or proton abstraction or and transfer.[16-19] Normally the charges are distributed evenly over the molecule which imparts neutral character to the molecule. Whenever the substitutions are added to the molecule, the charge distribution is completely altered with respect to the substitution or the removal of any atom/moiety. Here in the case of receptor R, the negative Mulliken charges populated over the highly electronegative atoms (N-atoms) which splits the positive charges among hydrogen atoms. Hence the N and H of the molecule have negative and positive space respectively. The Mulliken charges of **N** and **H** atoms of amine ($-NH_2$) of **R** and $\mathbf{R^-}$ are presented in table 1.

Table 1: Mulliken charges of N and H atoms of NH_2

Atom	R	R^-
N	-0.406	0.295
H	0.252	0.267

4. UV-Vis Spectra & Molecular Orbital Analysis

The UV–Vis spectra and the electronic transitions of the receptor molecule and its deprotonated form were calculated using the TD-DFT/6-311++G (d, p) method. Here F^- ion induced deprotonation (Fig. 1) resulted in increase of electron density on the nitrile group of ICT channel (D-π-A) in LUMO, while as there is a decrease in electron density in HOMO. The alteration clearly indicate Intramolecular charge transfer (ICT) enhancement.

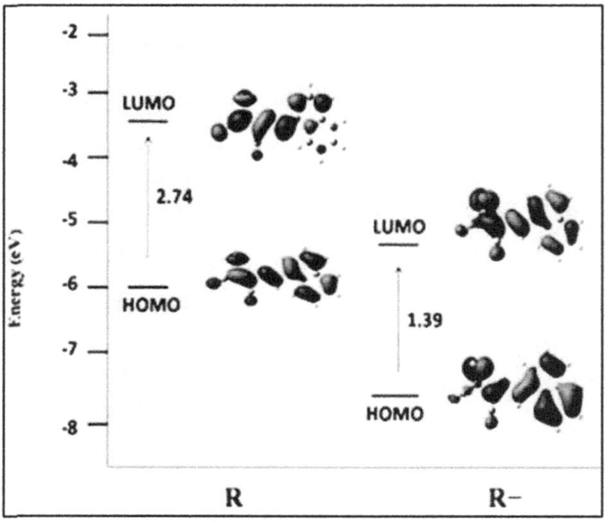

Fig 1. Excitation energies and HOMO and LUMO transitions of receptors before and after fluoride ion recognition R and R^-.

The ICT enhancement further results in the decrease of electron density over naphthyl moiety ring in the LUMO of R^- compared to its HOMO. This clearly indicates blockade of PET channel in the molecule after deprotonation. Hence fluorescence "turn on" could be understood in the molecule after fluoride addition due to ICT enhancement.

The mechanism of recognition can be explained in terms of the strong basicity of F^- in the non-aqueous environments (DMSO, THF, CH_3CN, etc). Under these conditions, F^- interacts with polarized –N–H moiety of amine group in the receptor molecule. The recognition is accompanied by the transfer of proton from $-NH_2$ to the F^- (Fig 2). The recognition is further driven by the formation of the thermodynamically stable HF_2^- dimer which was reported by author during the course of the work[20-21].

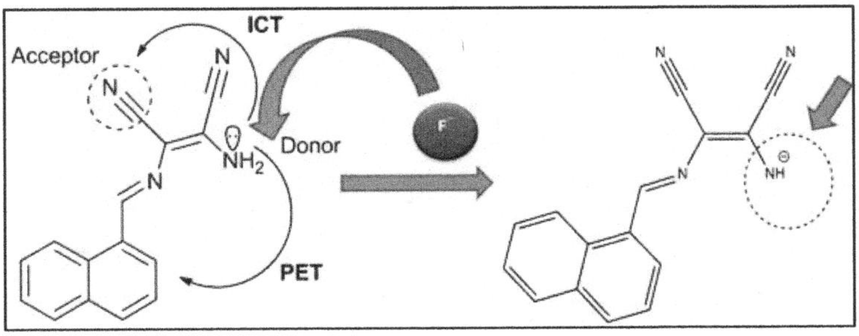

Fig. 2. Proposed mechanism of F^- interaction with the molecular receptor (R) leading to the formation of deprotonated receptor R^-.

Due to this deprotonation, receptor R in the presence of F^- displayed a red shift both experimentally as well as theoretically. In this study, the trends in spectral shift are as per experimental observations, but not empirically close with experimental (Fig. 4 and Table 2). This may be because of solvent effect and the nature of interaction

Table 2. Comparative experimental ad theoretical electronic absorption spectra of R and **R**$^-$

Receptor	**Experimental**			**Theoretical**		
R	385 nm	460 nm	75 nm	487.84 nm	820.31 nm	332.47 nm

The electron delocalization overall leads to the decrease in the energy gap between HOMO and LUMO of molecule from 2.74 eV to 1.39 eV (Fig 3a and 3b). Hence the presence of fluoride anion realizes spectral shift (red-shift), thus offering naked-eye optical detection off.

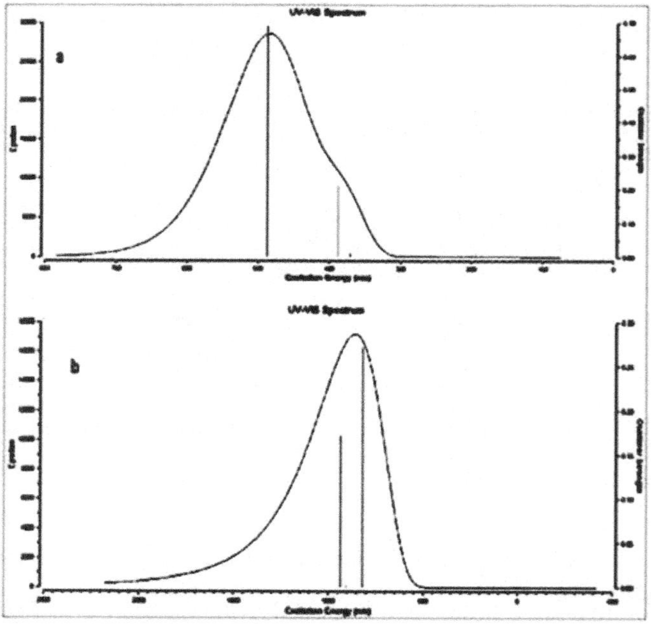

Fig. 3 Theoretically calculated electronic absorption spectra of **R** (a) and R$^-$ (b).

5. Conclusions

In summary, for the first time theoretically study of the structural and spectral aspects of a naphthyl moiety based molecular receptor tethered with DAMN fragment. For understanding fluoride recognition mechanism and the nature of bonds in the receptor, negative and positive electronic regions were revealed on N and H atoms of the molecule with the help of Mullikan charge distribution analysis. The theoretical calculations (DFT/TD-DFT) of the receptor displayed red shift of absorption spectra after interaction with the F^-. The reduction in the energy gap (ΔE) between highest occupied and lowest unoccupied energy levels were also revealed, hence strong the red shift of its spectral characteristics. Most importantly, during the course of proton transfer of $-NH_2$ to F^-, a decrease in the pi-electron density over napthyl moiety in the LUMO contrary to its HOMO revealed blockade of PET process and hence fluorescent turn on response.

6. References

1. M. A. Kaloo and J. Sankar. A molecular Boolean mimic with OR, NOR, YES and IN functions: dual-ion recognition driven fluorescence ''turn on'' (2014) New J. Chem. 38: 923.
2. Fluoride H.N. Kim, W.X. Ren, J.S. Kim and J. Yoon. Fluorescent and colorimetric sensors for detection of lead, cadmium and mercury ions (2012) Chem Soc Rev 41: 32210.
3. B. E. Kim, T. Nevit and D. J. Thiele. Mechanisms for copper acquisition, distribution and regulation. (2008) Nat Chem Biol 4: 176.
4. R. Uauy, M. Olivares and M. Gonzalez. Essentiality of copper in humans (1998) Am J Clin Nutr 67: 952.
5. D Karak, S Lohar, A Banerjee, A Sahana, I Hauli,et al. Interaction of soft donor sites with a hard metal ion: crystallographically characterized blue emitting fluorescent probe for Al (III) with cell staining studies (2012) RSC Adv 2: 12447.
6. P. de Silva, D. B. Fox, A. J. M. Huxley and T. S. Moody, Coord. Progress in Heterocyclic Chemistry: A Critical Review of the 2000 literature (2000) Chem Rev 205: 41.
7. R. Martınez, A. Espinoosa, A. Ta raga and P. Molina. Bis (indolyl) methane derivatives as highly selective colorimetric and ratiometric fluorescent molecular chemosensors for Cu^{2+} cations (2008) Tetrahedron 644: 2184.
8. AP de Silva, AJM Huxley and TS Moody. Combining luminescence, coordination and electron transfer for signaling purposes (2000). Coord Chem Rev 205: 41.
9. V. Amendola, L. Fabbrizzi. Anion receptors that contain metals as structural units (2009) Chem Commun 513.

10. A. Ajayaghosh, P Carool, S Sreejith. A Ratiometric Fluorescence Probe for Selective Visual Sensing of Zn^{2+} (2005)) J Am Chem Soc 127: 14962.
11. D Wang, Y Ke, H Guoo, J Chen, W Weng. A novel highly selective colorimetric sensor for aluminum (III) ion using Schiff base derivative (2014) Spectrochim. Acta, Part A 122: 268.
12. V Amendola, L Fabbriizzi, M Licchelli, C Mangano, P Pallavicini, Et al. Molecular events switched by transition metals (2000) Coord Chem Rev 192: 649.
13. AP Silva, HNQ Gunaratne, T Gunlaugsson, AJM Huxley, CP Mc Coy, et al. Signaling Recognition Events with Fluorescent Sensors and Switches (1997) Chem Rev 97: 1515.
14. P. D. Beer and P. A. Gale. Anion Recognition and Sensing: The State of the Art and Future Perspectives (2001) Angew Chem 40: 486.
15. C. Perez-Casas and A. K. Yatsimirsky, J. Detailing Hydrogen Bonding and Deprotonation Equilibria between Anions and Urea/Thiourea Derivatives (2008) Org Chem 73: 2275.
16. FY Wu, ZC Wen, N Zhou, YF Zhao, YB Jiang. A novel thiourea based dual fluorescent anion receptor with a rigid hydrazine spacer (2002) Org Lett 4: 3203.
17. J. M. Linares, D. Poweell and J. K. Bowman, Coord. Chem. Rev 20003, 240, 57.
18. X. J. Peng, Y. K. Wu, J. L. Fan, M. Z. Tian and K. L. Han, J. Org. Chem., 2005, 70, 10524.
19. Chellappan, N. J. Singh, I. C. Hwang, J. W. Lee and K. S. Kim. A calyx [4] imidazolium [2] pyridine as an anion receptor (2005) Angew Chem 44: 2899.

20. L. Yang, X. Li, J. Wanng, Y. Qu and J. Hua. Colorimetric and Ratiometric Near-Infrared Fluorescent Cyanide Chemodosimeter Based on Phenazine Derivatives (20013) ACS Appl Mater Interface 5: 1317.

21. J Ren, W Zhu and H Tian. A highly sensitive and selective chemosensor for cyanide (2008) Talanta 75: 760-764.

Unit-4

Phenyl-Diaminomalenonitrile Schiff base derivatives (Pages: 35-45)

1. Molecular Receptors of Interest

The design and synthesis of receptors capable of binding and sensing anions selectively have received significant attention in recent years because anions play an important role in a wide range of biological, environmental, and chemical processes A. Bianchi et al[1]. Especially, colorimetric and ratiometric chemo sensors Y. Deng et al appear to be particularly attractive due to their simplicity, high sensitivity, and high selectivity J.F. Callan et al[2-8]. For quantitative analyses, ratiometric chemo sensors have the significant advantage of their dual emission system, which can minimize measurement errors caused by fluctuations of light scattering as well as reagent concentration B. Valeur et al[9]. Colorimetric chemo sensors provide an immediate qualitative signal, which allows direct naked-eye detection of anions because of a specific color change of solution upon anion complexation by C. Coll et al[10-11]. Among the range of biologically important anions, fluoride, the smallest anion, has unique chemical properties. It is of particular interest to recognize and detect it owing to its essential role in a broad range of biological, medical, and chemical processes of osteoporosis, fluorination of drinking water supplies, or even in chemical and nuclear warfare agents K.L. Kirk et al[12-14]. However, the chemo sensors for F^- are rather few, and there is a need for good anion sensors with an improved specific response, in particular, with selectivity for F^- in the presence of other anions. Furthermore, it is necessary to understand the unique host-guest interaction of the sensor with F^- and other anions. Different signaling mechanisms have been suggested for F^-, such as photo induced electron transfer (PET) S.K. Kim et al, excited state proton transfer (ESPT) X. Peng et al, intramolecular charge transfer (ICT) Z. Li et al [15-18].

In the present study, absorption spectroscopic analysis of DAMN-based molecular receptors/ sensors 1, 2, 3 and 4 (Figure. 1) and their interaction with F^- ions has been carried out theoretically and discussed in detail. The combined use of DFT/TD-DFT (B3LYP) functional and standard basis sets 6-311++G (*d,p*) provides an excellent balance between accuracy and computational efficiency of absorption spectra. The literature survey reveals that, to the best of our knowledge, no intensive observation of theoretical [DFT/TD-DFT] investigation has been reported so far with this type of chemo sensor design possessing free $-NH_2$ based recognition site. Therefore, the present investigation was undertaken to study the absorption spectra and inter and intra molecular interaction between HOMO and LUMO energy levels of the mentioned chemo sensors.

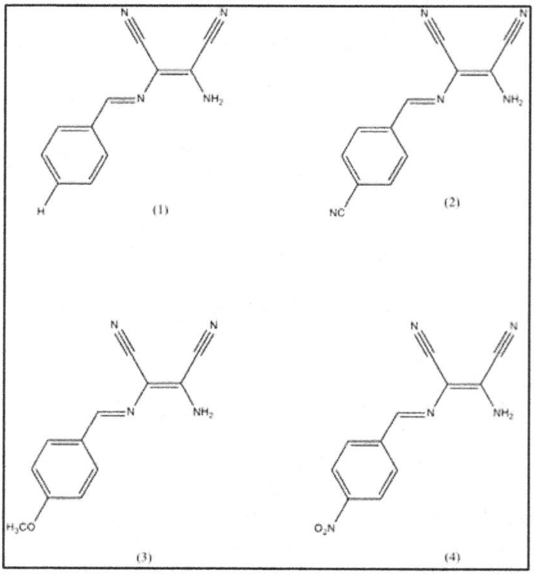

Fig 1. Generalized structure of the molecular receptor derivatives corresponding to the substituents at the para-position.

2. Computational Method

In the present work, one of the hybrid methods; B3LYP was carried out using the basis sets 6-311++G(d,p) using GAUSSIAN 09W program. In DFT methods; Becke's hybrid function combined with the Lee-Yang-Parr correlation function (B3LYP), predict the best results of Mullikan charge density on N and H of amine (NH_2) group and absorption spectra of DMN based Chemo sensors 1, 2, 3 and 4. The theoretical result of electronic absorption transitions was compared with the experimental [9-11].

3. Muliken Charge distribution analysis

The Mulliken charge on the atoms is used to understand the charge distribution on the chemical bonding because it facilitates positive and negative regions in the molecular space, at which the protons and electrons concentrate. Thus chemically significant regions; bonds can be identified; also gives the narration of the mechanism of electrophilic and nucleophilic substitutions. Normally, the charges are distributed evenly over the molecule which leads to be neutral. Whenever the substitutions are added to the molecule, the charge distribution is completely altered with respect to the substitution. Usually, the negative Mulliken populated over the highly electronegative atoms which splits the positive charges among hydrogen atoms. The N and H of the molecule have negative and positive space respectively. The Mulliken charges of N and H atoms of amine (NH_2) of chemo sensors are presented in **Table 1.**

Table 1. Mulliken charges of N and H atoms of NH_2

Atom	**1**	**d-1**	**2**	**d-2**	**3**	**d-3**	**4**	**d-4**
N	-0.779	-0.306	-0.777	-0.276	-0.492	-0.337	-0.490	-0.276
H	0.348	0.216	0.339	0.220	0.243	0.247	0.252	0.220

4. UV-Vis Spectra & Molecular Orbital Analysis

The UV–Vis spectra and the electronic transitions of the molecular receptors or sensors have been calculated using the TD-DFT/6-311++G (d,p) method and has been compared with experimental as shown in **Table 2, Figure 2.** The local HOMO electrons delocalized through phenyl moiety and amine group. In the local transition, these delocalized electrons (HOMO) were transited only to phenyl moiety (LUMO). The delocalization is more in case of electron-withdrawing substituent's at *para*-position. Thus the decrease in the energy gap between HOMO and LUMO (FIGURE 2) of receptor in presence of fluoride anion realizes spectral shift (red-shift), and hence offering naked-eye optical detection of anion.

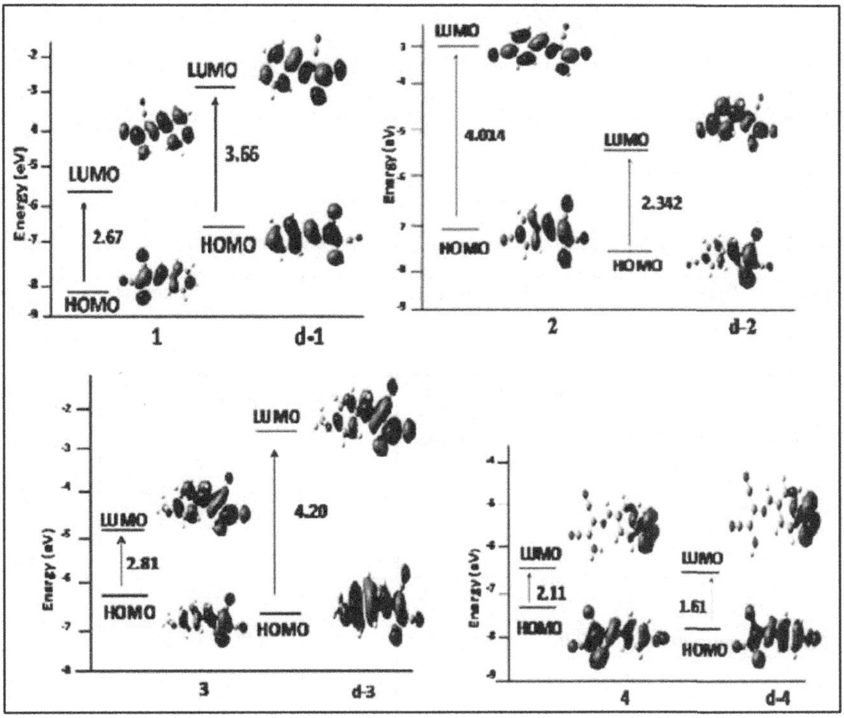

Figure 2. Excitation energies and HOMO LUMO transitions of various molecular receptors 1, 2, 3 and 4. The energy gap is in electron volt (eV).

The theoretical results are not good agreement with experimental may be because of solvent effect and the nature of interaction. The molecular receptors sensors undergo either H-bonding or deprotonation after the interaction with the F− ions depending upon the concentration of F- ions and the nature of phenyl-para substituent. The phenyl para-position plays an important role in the structure-selectivity relationship by tuning the acidity of the amine and facilitating the intramolecular charge transfer (ICT). The molecular receptors sensors 1, 2, 3 and 4 undergo deprotonation after the interaction with the F− ions. Fluoride-induced deprotonation d-1, d-2, d-3 and d-4 results in a charge transfer (CT) transition with red shift. The resulting negatively charged species with high charge density on the amine nitrogen is associated with the enhancement of the push–pull effect of the ICT transition. The energy gap and oscillation strength was also calculated to all DMN derivatives Table 2 & Table 3.

Table 2. Comparative experimental ad theoretical electronic absorption spectra of 1, 2, 3 and 4.

Molecular Receptor	**Experimental**			**Theoretical**		
1 (Receptor)	**λ (nm)**	**λabs F (nm)**	**Δλ (nm)**	**λ (nm)**	**λabs F (nm)**	**Δλ (nm)**
	363	422	59	335	516	181
2 (CN−derivative)	376	465	89	337	493	157
3 (OCH_3−derivative)	373	423	50	328	558	230
4 (NO2−derivative)	393	650	167	774	1240	466

Table 3. Energy gap and oscillation strength of Various Molecular Receptors 1, 2, 3 and 4.

Molecular Receptor	**E (eV)**	**Oscillation strength (*f*)**	**Composition**
1	2.67	0.509	
d-1	3.66	0.0336	
2	4.02	0.162	
d-2	2.34	0.0184	**HOMO → LUMO**
3	2.81	0.4805	
d-3	2.36	0.0576	
4	2.11	0.0033	
d-4	1.61	0.0004	

The receptors 2 and 4 owing to its strong electron withdrawing –CN and –NO_2 groups respectively make the charge delocalization feasible on the interaction with fluoride and is associated with a very high shifts 157 nm and 466 nm respectively in their absorption spectra (Figure 3). However the receptor 1 undergoes deprotonation because of high concentration of F– ions is reflected with shift of 181 nm in its absorption spectrum. On the other hand, the receptor 3 owing to its electron donating group −OCH3 is associated with 230 nm shift in the absorption spectrum because of deprotonation after interaction.

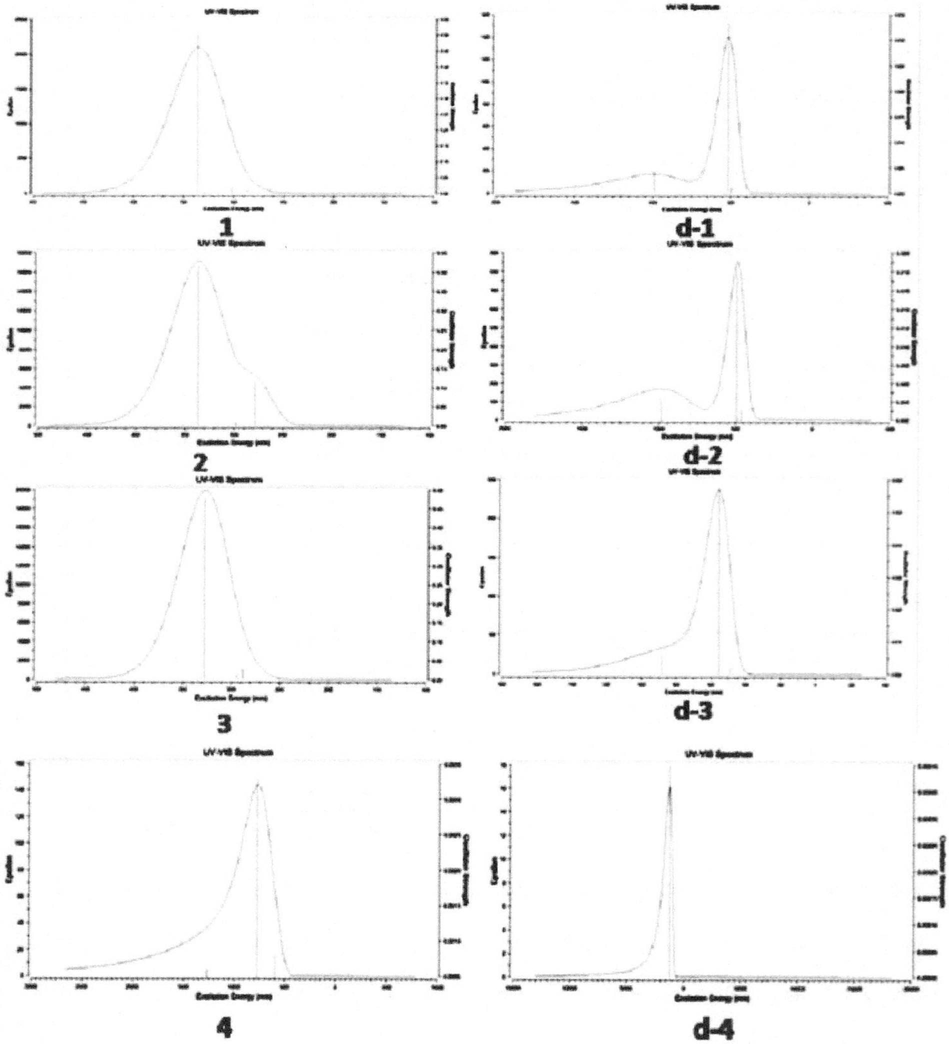

Figure 3. Theoretically calculated electronic absorption spectra of 1, 2, 3 and 4.

5. Tuning of electronic effects

In order to check the effect of various substituent's (from electron donating to the electron with-drawing substituent's) present on the phenyl ring, absorption and emission studies were carried out under similar conditions. The studies were carried out both in the presence as well as absence of the fluoride ion in the form of tetra butyl ammonium fluoride under the organic conditions. All the studies were carried out at the room temperature conditions. In the absorption spectrum, various receptor derivatives (-CN, $-NO_2$, -H and $-OCH_3$) displayed absorption maximum around 376 nm, 393nm, 363nm, and 373nm respectively. The extinction coefficient values have been further calculated at these characteristic absorption peaks for each of the receptor derivatives. All these values have been tabulated below (Table 4).

In the next step, absorption behavior of all these receptors was checked n the presence of fluoride ion. In the presence of fluoride ion, the absorption maximum was seen to be shifted towards higher wavelengths (red-shift). The absorption peak of each receptor derivative shifted to 465nm, 650nm, 422 nm and 423 nm respectively. The value of the red-shift was further calculated. The values are present in the table 4.

Following the absorption spectral derivatives, emission spectral data was collected for all these molecular receptors. It was observed that each receptor displayed a characteristic emission peak at 404 nm, 501 nm, 428 nm and 438 nm respectively upon excitation at their characteristic absorption maximum.

Upon addition of fluoride ion all these receptors were further monitored via the emission spectroscopy. It was observed that their chacteristic emission spectra got further shifted towards higher wave lengths. The new emission peaks were observed at 404 nm, 485 nm, 425 nm and 436 nm respectively. The completed details have been mentioned in the table below (Table 4).

Table 4. Experimental absorption spectral and emission spectral details of various Molecular Receptors.

Receptor	λ_{max} (nm)	ε ($Lmol^{-1}cm^{-1}$)	λ_{max} F (nm)	Δλ	λ_{em} (nm)	λ_{em} F(nm)
-CN	376	33160	465	89	404 (*w*), 426(*m*)	404 (*h*), 423(*h*)
$-NO_2$	393	34894	650	167	501 (*w*)	485 (*w*)
-H	363	20841	422	59	428 (*w*)	425(*w*)
$-OCH_3$	373	43885	423	50	438 (*w*)	436 (*w*)

Besides all the above, the binding behavior was further confirmed with the help of jobs studies. It is used to determine the stoichiometry of a binding event. Different concentration of fluoride ion and the receptor were prepared. These were later monitored by the absorption spectroscopy at their absorption peaks. The job's plot for the fluoride ion and cyanide receptor derivative is shown below (Fig. 4). It can be cleared seen that the stoichiometry of the fluoride ion and receptor interaction is clearly 1: 1.

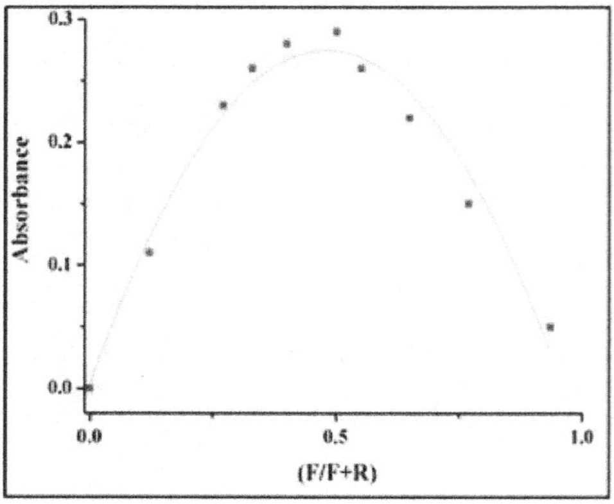

Fig 4. Jobs plot for fluoride ion and receptor binding (cyanide derivative receptor).

Following the jobs studies, in order to check the sensitivity of the binding events of various receptor derivatives for fluoride ion, binding constants were calculated for each receptor ion in presence of fluoride ion in the organic solvents at room temperature.

The binding constants (K, M^{-1}) of receptors with various fluoride anion were calculated from absorption studies using the Benesi-Hilderbrand equation[19-20]:

$$\frac{1}{(A-A')} = \frac{1}{(A''-A')} + \left[\frac{1}{K}\right] 1/(A''-A')$$

where **A** is the absorbance of compound before anion addition, **A′** for the absorbance in presence of analyte, **A″** refers to the absorption upon saturation with fluoride anion and **K** indicates the binding constant. Upon plotting inverse of change in absorbance of newly emerged absorption signal after analyte addition ($1/\Delta A = 1/A-A'$), with respect to inverse of analyte concentration ($1/F^-$). Binding constant (K) was determined from the ratio of intercept and slope of the curve.

From the binding constant values (Table 5), it is clear that the highly electron withdrawing substituent results in the strong change in the ion recognition of fluoride ion. Here one can say more id the binding constant more efficient is the recognition vent and hence more is the sensitivity.

Table 5. Table representing the binding content value of various receptor derivatives.

Receptor	Binding constant (K, M^{-1})
-CN	$5.36*10^4 \pm 3495.13$
$-NO_2$	$8.58*10^3 \pm 976.83$
-H	$1.30*10^3 \pm 809.74$
$-OCH_3$	$1.70*10^2 \pm 138.66$

6. Effect of Aromatic Framework

In addition to the above engineering on the receptor frame work regarding fluoride ion recognition, tuning of electronic effects further checked by bringing in the various aromatic derivatives. Various aromatic derivatives were tethered in place of the phenyl ring (Figure 5). Among the various aromatic groups which were bought in and checked for fluoride recognition events include cinamyl, -phenyl, ferrocene, napthyl and pyrene moiety. The graph displaying the sensitivity has been shown below in the figure 5. It can be predicted that the sensitivity of the molecular receptor derivatives increases with increase in the aromaticity. Hence one can say that despite the fact that, we have electron withdrawing natured substituent on the other side of receptor frame work, aromatic substitutions made dramatic change in the sensing ability of the receptor towards anions.

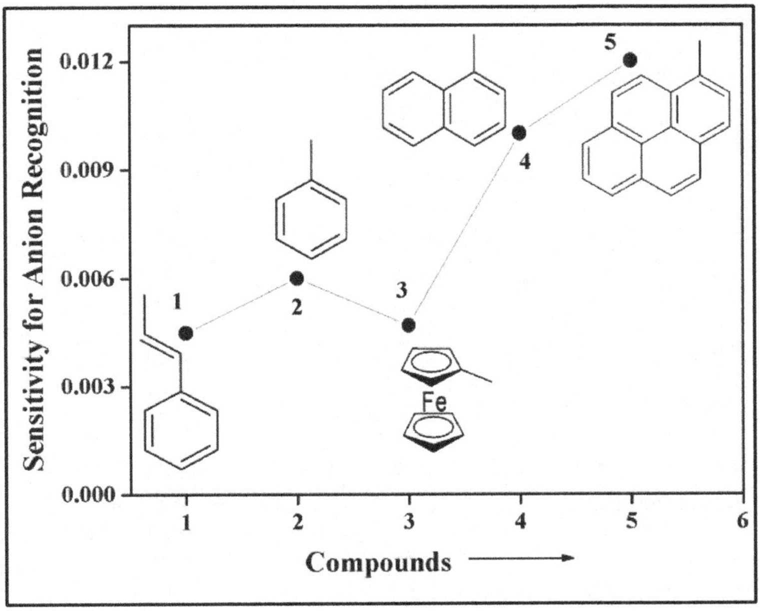

Fig 5. Sensitivity displayed by various aromatically tuned ion receptors for fluoride ion recognition. The details have been checked at room temperature.

In order to check the effect of these aromatic substituent's (from pheny ring to the pyrene ring) UV-Vis absorption studies were carried out under similar conditions as was the case earlier. The studies were carried out both in the presence as well as absence of the fluoride ion in the form of tetra butyl ammonium fluoride under the organic conditions. All the studies were carried out at the room temperature conditions. In the absorption spectrum, various receptor derivatives (cinamyl, -phenyl, ferrocene, napthyl and pyrene named as 1, 2, 3, 4 and 5 respectively) displayed absorption maximum around 376 nm, 363nm, 366nm, 390 nm and 422 nm respectively. All these values have been tabulated below (Table 6).

Following this, absorption studies were monitored in presence of fluoride ion as well under the similar conditions. In all these case, the absorption peak of the receptor derivative got red shifted to 455 nm, 422 nm, 406 nm, 460 nm, 517 nm for 1, 2, 3, 4 and 5 respectively. Among all these pyrene has shown larger shift in the absorption peak while as ferrocene has shown the least. This can be attributed to the aromaticity. As pyrene has large extended aromatic frame work, it has displayed large sensitivity as well as red shift in presence of the fluoride ion.

Table 5. Table representing the absorption spectral data of various receptor derivatives in presence of fluoride ion.

Receptor	A_1(nm)	A_2(nm)	A-F (nm)
1	376	0	455
2	363	0	422
3	366	496	406
4	390	0	460
5	422	0	517

4. Conclusions

In this work, various structural aspects and UV-visible absorption spectral characteristics of novel group of receptors recently proposed from DAMN molecular framework were studied by computational methods. The theoretical calculations (DFT/TD-DFT) of various chemo sensors displayed red shift of absorption spectra after the interaction with the F^- ion. The reduction in the energy gap (ΔE) between highest occupied and lowest unoccupied energy levels was revealed, leading to strong the red shifted of its absorption characteristics. The extent of red-shift was further understood *via* Mulliken charge distribution analysis, which signifies polarized nature of N-

H bond in free $-NH_2$ group in different derivatives. The large shift in the absorption spectra in comparison to the calculated ones may be due to the solvent effect and nature of interaction between the Receptor and F^- ion.

6. References

1. Bianchi A, Bowman JK, Garcia EE Supramolecular Chemistry of Anions. Wiley-VCH: New York, NY, USA, 1997.
2. Deng Y, Chen Y, Cao D, et al. A cationic triarylborane as water-tolerant Fluorescent chemosensor for fluoride anions. Sensors Actuator B-Chem. 2010; 149:165-169.
3. Li JQ, Li XY. Multichannel photo induced intramolecular electron-transfer excitations in a bis-naphthalimide spermine conjugate by time-dependent density functional theory. Phys Chem A 2007; 111:1306-13068.
4. Hagihara S, Tanaka H, Matile S. Boronic acid converters for reactive hydrazide amplifiers: Polyphenol sensing in green tea with synthetic pores. J Am Chem Soc 2008; 130:5656-5657.
5. Yoon J, Kim SK, Singh, et al. Imidazolium receptors for the recognition of anions. Chem Soc Rev 2006; 35: 355-360.
6. Gale PA. Structural and molecular recognition studies with acyclic anion receptors. Acc Chem Res 2006; 39:465-475.
7. Callan JF, de Silva AP, Magri DC. Luminescent sensors and switches in the early 21st century. Tetrahedron 2005; 61: 8551-8588.
8. Martinez MR, Sancanon F. Fluorogenic and chromogenic chemosensors and reagents for anions. Chem Rev 2003; 103:4419-4476.
9. Valeur B Molecular Fluorescence Wiley-VCH: Weinhein, Germany, 2002.

10. Coll C, Marínez MR, Dolores MM, et al. A simple approach for the selective and sensitive colorimetric detection of anionic surfactants in water. Angew Chem Int Ed 2007; 46; 1675-1678.

11. Evans LS, Gale PA, Light ME, et al. Anion binding vs deprotonation in colorimetric pyrrolylamidothiourea based anion sensors. Chem Commun 2006; 55:965-967.

12. Kirk KL. Biochemistry of the Halogens and Inorganic Halides; Plenum Press: New York, NY, USA, 1991.

13. Kleerekoper M. The role of calcitonin in the prevention of osteoporosis. Endocrin Metab Clin North Am. 1998; 27:414-418.

14. Zhang SW, Swager TM. Fluorescent detection of chemical warfare agents: Functional group specific ratiometric chemosensors. J Am Chem Soc 2003; 125:3420-3421.

15. Kim SK, Yoon J. A new fluorescent PET chemosensor for fluoride ions. Chem Commun 2002; 33:770-771.

16. Peng X, Wu Y, Fan J, et al. Colorimetric and ratiometric fluorescence sensing of fluoride: Tuning selectivity in proton transfer. J Org Chem 2005; 70:10524-10531.

17. Li Z and Zhang J An efficient theoretical study on host-guest interactions of a fluoride chemosensor with F^-, Cl^-, and Br^-. Chem Phys 2006; 331:159-163.

18. Thiagarajan V and Ramamurthy P Specific optical signalling of anions via intramolecular charge transfer pathway based on acridinedione fluorophore. J Lumin 2007; 126:886-892.

19. (a) Polster, J.; Lachmann, H. Spectrometric Titrations. VCH: Weinheim, **1989**.

20. Connors, K. A. Binding Constants. The Measurement of Molecular Complex. Wiley: New York, **1987**. (c) Benesi, H. A.; Hildebrand, J. H. J. Am. Chem. Soc. **1949**, *71*, 2703.

Publisher: Eliva Press Global Ltd

Email: info@elivapress.com

Made in United States
Orlando, FL
04 November 2026

90201142R00039